DISCARD

# AH-1W SUPER COBRAS

BY CARLOS ALVAREZ

BELLWETHER MEDIA · MINNEAPOLIS, MN

Are you ready to take it to the extreme? Torque books thrust you into the action-packed world of sports, vehicles, and adventure. These books may include dirt, smoke, fire, and dangerous stunts. WARNING: read at your own risk.

Library of Congress Cataloging-in-Publication Data

Alvarez, Carlos, 1968-
AH-1W Super Cobras / by Carlos Alvarez.
p. cm. – (Torque: Military machines)
Includes bibliographical references and index.
Summary: "Amazing photography accompanies engaging information about AH-1W Super Cobras. The combination of high-interest subject matter and light text is intended for students in grades 3 through 7"–Provided by publisher.
ISBN 978-1-60014-494-3 (hardcover : alk. paper)
1. Huey Cobra (Helicopter)–Juvenile literature. I. Title.
UG1233.A459 2010
623.74'63–dc22 2010000865

This edition first published in 2011 by Bellwether Media, Inc.

No part of this publication may be reproduced in whole or in part without written permission of the publisher. For information regarding permission, write to Bellwether Media, Inc., Attention: Permissions Department, 5357 Penn Avenue South, Minneapolis, MN 55419.

Text copyright © 2011 by Bellwether Media, Inc. TORQUE and associated logos are trademarks and/or registered trademarks of Bellwether Media, Inc.

The images in this book are reproduced through the courtesy of: Ted Carlson/Fotodynamics, front cover, pp. 7, 8-9, 10-11, 14, 17, 19, 20 (top); all other photos courtesy of the United States Department of Defense.

Printed in the United States of America, North Mankato, MN.
010111 1183

# CONTENTS

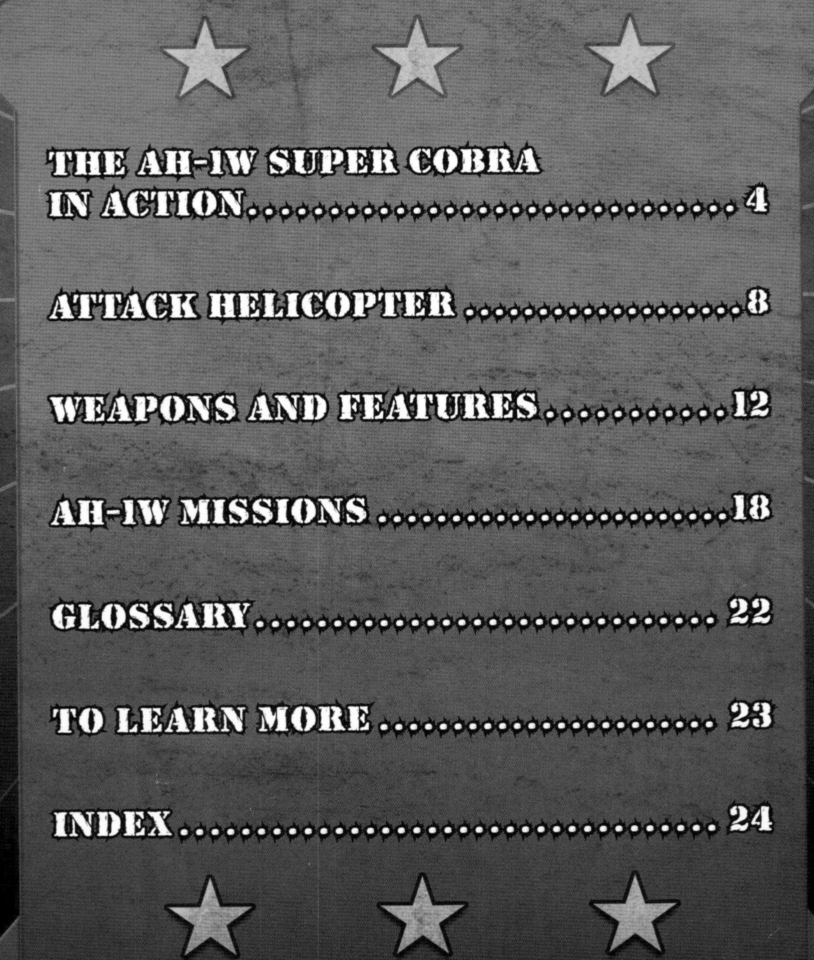

# THE AH-1W SUPER COBRA IN ACTION

A unit of the United States Marine Corps is moving across the desert. The Marines are headed toward an enemy base. The enemy knows they are coming. Several enemy tanks move into position to defend the base. They aim their huge guns at the Marines. Two AH-1W Super Cobras fly over the Marines.

MARINES
31

The Super Cobras head straight for the enemy tanks. They fire powerful AGM-114 Hellfire **missiles**. The missiles blast through the thick tank **armor**. Flaming debris flies through the air. The tanks are destroyed. The Marines are safe. The Super Cobras return to base.

# ATTACK HELICOPTER

The AH-1W Super Cobra is a powerful attack helicopter. It is used by the U.S. Marine Corps. The Super Cobra's sensors and weapons make it a dominant force in any battle. It can perform **missions** during the day or at night. It is useful over land or at sea. It can even fly in bad weather.

MARINES
33
160822
MARINES

AH-1 Cobra
MARINES
10
33
MARINES
23
RESCUE

The AH-1W is based on the United States Army's AH-1 Cobra. The Cobra was an attack helicopter used during the Vietnam War. The U.S. Marine Corps improved the Cobra design. They added a second engine and more weapons. The result was the AH-1W Super Cobra. The AH-1W entered U.S. Marine Corps service in 1986.

AH-1W Super Cobra

# WEAPONS AND FEATURES

The Super Cobra carries a wide range of weapons. A **turret** holds a 20mm cannon. Two weapons stations are on each wing of the helicopter. Marines can mount missiles, bombs, or Hydra 70 **rockets** to these stations.

Hydra 70 rocket

**★ FAST FACT ★**

The 20mm cannon can fire 650 rounds per minute.

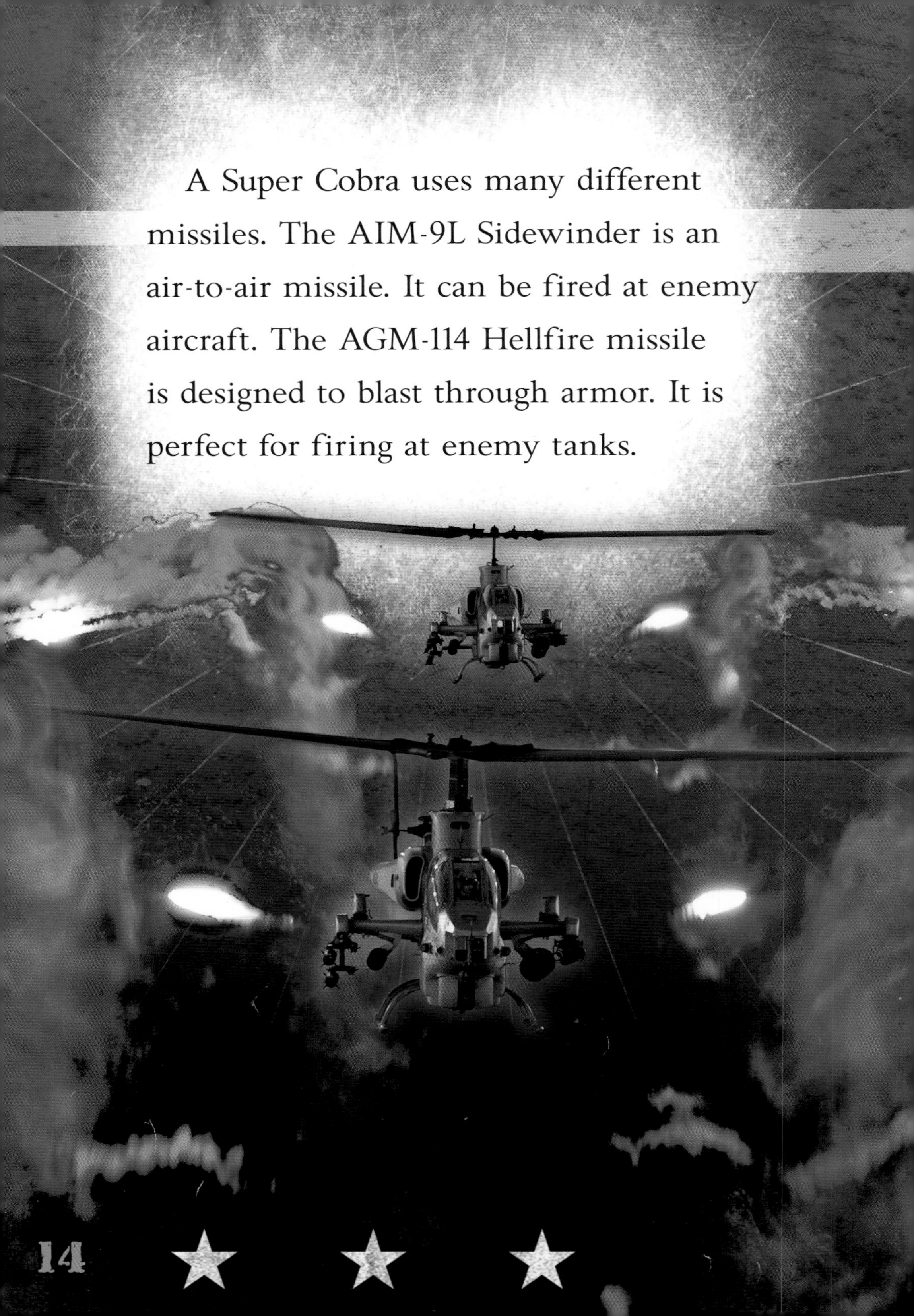

A Super Cobra uses many different missiles. The AIM-9L Sidewinder is an air-to-air missile. It can be fired at enemy aircraft. The AGM-114 Hellfire missile is designed to blast through armor. It is perfect for firing at enemy tanks.

★ FAST FACT ★

The Super Cobra uses countermeasures to protect itself from enemy missiles.

The **night targeting system (NTS)** helps Super Cobra crews select and fire at targets. The NTS finds and locks on to the heat of objects. This system and the firepower of a Super Cobra make the AH-1W a deadly foe.

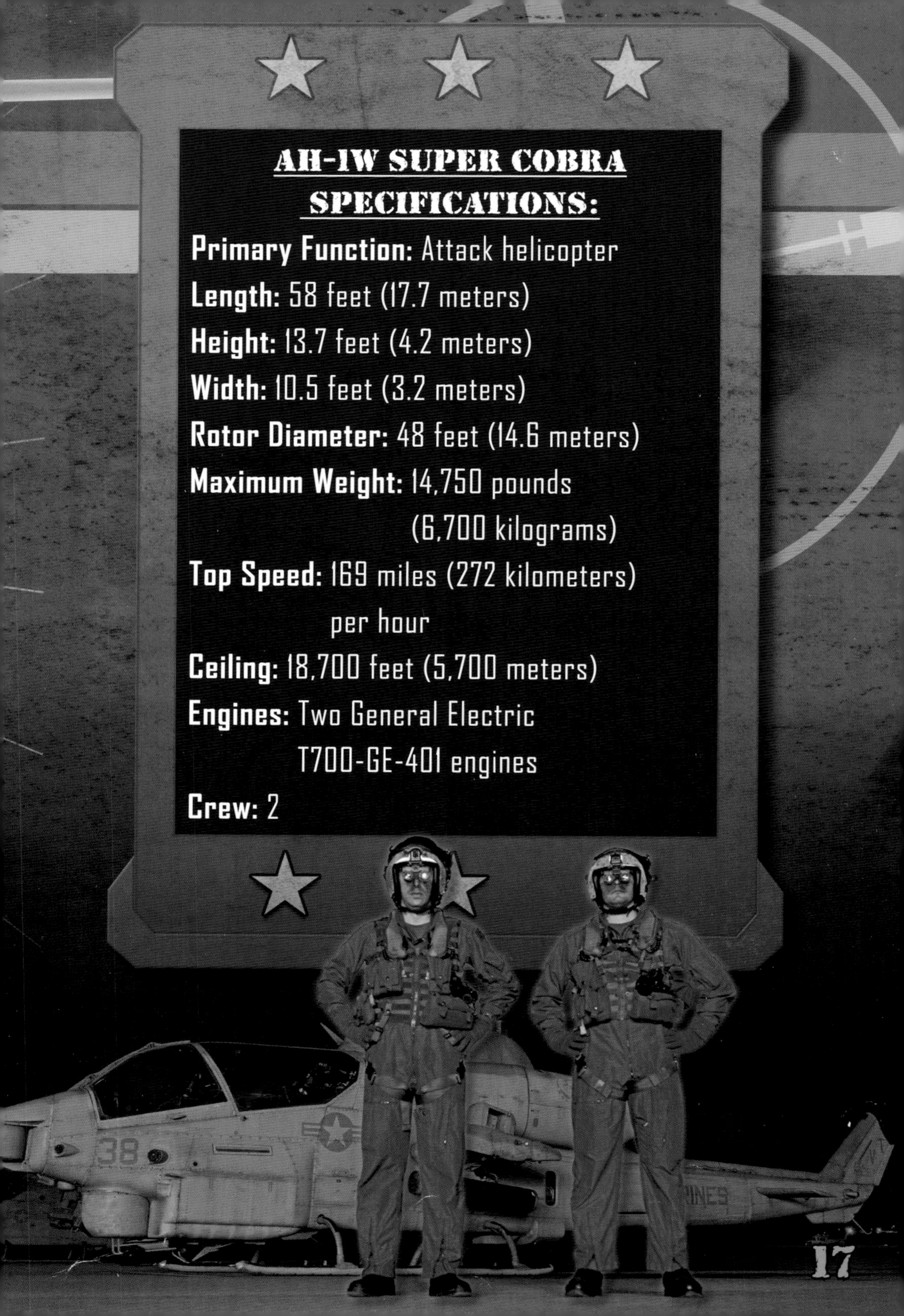

## AH-1W SUPER COBRA SPECIFICATIONS:

**Primary Function:** Attack helicopter

**Length:** 58 feet (17.7 meters)

**Height:** 13.7 feet (4.2 meters)

**Width:** 10.5 feet (3.2 meters)

**Rotor Diameter:** 48 feet (14.6 meters)

**Maximum Weight:** 14,750 pounds (6,700 kilograms)

**Top Speed:** 169 miles (272 kilometers) per hour

**Ceiling:** 18,700 feet (5,700 meters)

**Engines:** Two General Electric T700-GE-401 engines

**Crew:** 2

# AH-1W MISSIONS

The Super Cobra can perform many kinds of missions. It can attack enemy planes, helicopters, and ground vehicles. It can serve as an **escort** for transport helicopters. It can use a laser to mark targets for bombers. It can even go on **reconnaissance** missions.

★ FAST FACT ★

The Super Cobra is very narrow. This makes it difficult to see, and hard to shoot down.

23

The Super Cobra carries a pilot and a **co-pilot gunner (CPG)**. The pilot sits behind the CPG. The pilot's seat is raised so the pilot can see over the CPG. The pilot flies the helicopter. The CPG operates the weapons.

The Super Cobra performs many dangerous missions. Skilled pilots, advanced technology, and powerful weapons have made the Super Cobra a valuable tool for the U.S. Marine Corps.

The U.S. Marine Corps plans to replace the Super Cobra with the AH-1Z Viper sometime in the 2010s.

# GLOSSARY

**armor**—protective plating

**co-pilot gunner (CPG)**—the crew member who operates a Super Cobra's weapons and helps the pilot fly the helicopter

**escort**—to travel alongside and protect

**missiles**—explosives launched at targets on the ground or in the air

**missions**—military tasks

**night targeting system (NTS)**—a computer that helps an AH-1W crew find targets and fire weapons

**reconnaissance**—secret observation

**rockets**—flying explosives that are not guided

**turret**—a weapon mount that can rotate in any direction; a Super Cobra's 20mm cannon is mounted on a turret.

# TO LEARN MORE

## AT THE LIBRARY

David, Jack. *United States Marine Corps.* Minneapolis, Minn.: Bellwether Media, 2008.

Green, Michael, and Gladys Green. *Super Cobra Attack Helicopters: The AH-1W.* Mankato, Minn.: Capstone Press, 2005.

Kaelberer, Angie Peterson. *U.S. Marine Corps Assault Vehicles.* Mankato, Minn.: Capstone Press, 2007.

## ON THE WEB

Learning more about military machines is as easy as 1, 2, 3.

1. Go to www.factsurfer.com.

2. Enter "military machines" into the search box.

3. Click the "Surf" button and you will see a list of related Web sites.

With factsurfer.com, finding more information is just a click away.

# INDEX